HERB LIBRARY

CAMOMILE

Kate Ferry-Swainson has been a writer and editor for fifteen years, specializing in gardening, history, mythology, and self help. As a writer she has most recently contributed extensively to *Mindpower*, a major series of self-help books; has co-written *Spirit Stones*, a book about Native American spirituality, and has written the historical section for the art historical monograph *Van Dyck*. She has also written *Herb Library: Mint* and *Herb Library: Ginger* in the same series as this volume.

Deni Bown is a freelance writer and photographer specialising in botany, gardening, herbs, and natural history. The author and photographer of *The RHS Encyclopedia of Herbs and Their Use*, her other books include *Growing Herbs* and *RHS Plant Guides: Garden Herbs*. As President of The Herb Society, she lectures on various aspects of herbs and other plant topics, and has travelled throughout the world in pursuit of her subject as well as conducting specialist garden tours. She is a regular researcher and photographer at both Kew and Edinburgh Botanical Gardens.

Dedication:
For Richard, Matilda and Wriggly Baby

THIS IS A CARLTON BOOK

Text and Design copyright © 1999 Carlton Books Limited

This edition published by
Carlton Books Limited 1999
20 Mortimer Street
London
W1N 7RD
www.carlton.com

A CIP catalogue for this book is available from the British Library.

ISBN 1 85868 695 4

Editorial Manager: Penny Simpson
Art Editor: Tim Brown
Design: blu inc graphics
Picture Research: Frances Vargo
Production: Alexia Turner

Printed in Dubai

HERB LIBRARY

CAMOMILE

Kate Ferry-Swainson

Series Editor: Deni Bown

CARLTON

CONTENTS

INTRODUCTION

It hath floures wonderfully shynynge yellow and resemblynge the appell of an eye ... the herbe may be called in English, golden floure. It will restore a man to hys color shortly uf a man after the longe use of the bathe drynke of it after he is come forthe out of the bathe. This herbe is scarce in Germany but in England it is so plenteous that it groweth not only in gardynes but also VIII mile above London, it groweth in the wylde felde, in Rychmonde grene, in Brantfurde grene ... Thys herbe was consecrated by the wyse men of Egypt unto the Sonne and was rekened to be the only remedy of all agues.

William Turner

Imagine yourself in your garden on a bright summer's morning.
You wander barefoot along your camomile path, enjoying the
feel of the soft, feathery foliage between your toes, savouring the
fresh apple perfume that wafts up to you from the plants beneath
your feet. Just the smell of it is calming.
You take a seat in a sunny spot, the sun
warming the plants alongside, sending the
aroma of the camomile flowers wafting
towards you. You inhale, sit back, feel your
senses embracing the scent and welcome
the calming aura that descends on you
from within.

The benefits experienced outdoors
from camomile's perfume inspire you to
bring it into your home – or even to bring
it into your body. Either way, the therapeutic effects are
multiplied: you may drink a tisane made from camomile flowers,
swish the flowers and leaves through your bath, light a candle
scented with camomile, or receive a massage with camomile oil.
Ancient Egyptians, ancient Greeks, Renaissance herbalists,
Elizabethan ladies and busy modern people have all experienced
the therapeutic benefits of camomile: revelling in the scent, and
employing the herb in ways that heal everyday ailments as well
as give pleasure to mind and body.

Camomile's superb pedigree as a therapeutic herb derives
from its mild, soothing and calming nature which may be
harnessed to treat a wide range of ailments including nervous
digestion, insomnia and minor wounds. Its mildness makes it
particularly suitable for use with babies and children.

This book shows you how to capture the scent, gentleness
and healing properties of camomile flowers and essential oil. The
first chapter examines the history and development of its use as a
therapeutic herb: a fantastic epic stretching from the tombs of

Relax awhile on
a seat planted
with a cushion of
non-flowering,
scented
camomile.

Opposite:
The flowers of
Matricaria recutita,
German
camomile.

the Egyptian Pharaohs 5,000 years ago to the present day. During the long, dark days of Anglo-Saxon and medieval Britain, the magical nature of the herb once again began to be analysed. The royal court of Elizabeth I launched the age of discovery when new lands and thus new plants were being discovered, and botanist-physicians raced to understand, categorize and illustrate the healing herbs. By the beginning of the nineteenth century, with the scientific revolution, herbal remedies had slipped out of the hands of ordinary people – the 'old wives' who had traditionally been responsible for collecting herbs, making preparations, prescribing and applying the remedies – to be lost in the tide of the modern drug industry. Now, however, a new scientific and popular interest is resurging.

Opposite: Walking along a scented camomile path is a sensual experience.

VARIETIES OF CAMOMILE

Camomile has been spelled thus and with an 'h', as chamomile, variously throughout history. Both spellings are correct although the Royal Horticultural Society and the *British Herbal Pharmacopoeia* now favour the latter spelling. There are two species of camomile used for herbal remedies.

Chamaemelum nobile, Roman camomile, is an evergreen perennial, and a cultivar *C. nobile* 'Flore Pleno' is double flowered; *Matricaria recutita*, German camomile, is an annual plant. The two species, both from the *Asteraceae* family, have very similar qualities and properties, but German camomile is much more pleasant to drink in tisanes than is Roman camomile, which is too bitter to drink for pleasure. Roman camomile is, however, better in medicinal remedies, although the German is slightly richer in essential oil and is more analgesic, which is particularly beneficial in the case of treating burns and preventing infection. In order to capitalize

The plentiful flowers of *Chamaemelum nobile* 'Flore pleno' are a delight to both apothecaries and gardeners.

9

The daisy-like flowers of German camomile.

Opposite: Camomile growing in abundance in the Ligurian Alps, Italy.

on the benefits of each species, herbalists often use a mixture of one part Roman to two parts German camomile. Other common names by which the plants are known include scented mayweed, corn feverfew, garden camomile, sweet camomile and true camomile. The cultivar *C. nobile* 'Treneague' does not flower and so is not used for herbal remedies and cosmetics although it is particularly suitable for creating lawns or paths and for cushioning raised seating areas.

Camomile has soft, feathery, green foliage and white daisy-like flowers with yellow centres. The flowers, borne on long stalks in summer, have an earthy appley scent, and the leaves are long and finely divided. Roman camomile forms a dense mat, growing to about 15 cm (6 in) high, with a spread of about 45 cm (18 in); German camomile reaches a height of up to 60 cm (2 ft), with a spread of up to 38 cm (15 in); its stems are multi-branched. The flowers of both species contain the aromatic, therapeutic compounds, and therefore the cultivar 'Flore Pleno' is of particular value because it produces twice the amount of flower. Double-flowered cultivars such as this often occur in the wild and are commonly taken into cultivation and propagated in favour of single-flowered plants.

Right: The non-flowering cultivar *Chamaemelum nobile* 'Treneague' is used for planting up paths and garden seats.

Both species of camomile are native to south and west Europe and the Mediterranean region. They are cultivated in Britain, Belgium, Hungary, the USA, Italy, France and Egypt. Roman camomile is grown mainly in France, Belgium and eastern Europe,

while German camomile is cultivated predominantly in
Argentina, Egypt and south-eastern Europe.

GROWING CAMOMILE

You may choose to grow your own plants in a herb garden,
mixed border or container. While in the past country houses
boasted extensive and complex herb gardens, all you need is a
small sunny corner, patio or balcony. Plants such as this make a
valuable addition to any outdoor living area, by providing scent:
on a warm, summer's day run your hand gently over the plant to
release the aroma.

Camomile is
a decorative
addition to a
garden, as a
lawn or planted
in an urn.

Chamaemelum nobile is an evergreen hardy perennial, meaning
that it can usually survive a British winter outside,
although in very cold or wet weather it may look
bedraggled; nevertheless, it is wise to take the
precaution of overwintering some offsets in case
the plants don't survive the winter. It likes to be
planted in light, well-drained soil and situated in a
sunny spot. You may buy young plants from a
garden centre or buy seeds, which should be
planted in spring or autumn. However, 'Flore
Pleno' and 'Treneague' are not grown from seed and need to be
propagated vegetatively.

Matricaria recutita is an annual plant, and is grown from seed
sown in spring or autumn. The plant will die at the end of the
growing season. It also likes to be planted in a sunny spot in
well-drained soil. This plant may grow quite large, although if it
is grown in a container its size may be controlled.

CULTIVATING CAMOMILE

There is very little routine maintenance to be done with either
species. You should certainly ensure that, if you are planting the
herb in the open garden, you leave enough space for it to grow

to its full size. Perhaps most importantly, however, ensure that during the flowering season you pick off dead flowers promptly: this spurs on the plant to produce more flowers. Water the plants thoroughly, particularly when they are becoming established, and remove weeds regularly to prevent the plants from becoming choked.

Camomile plants are readily available in pots from nurseries.

HARVESTING AND STORING FLOWERS

In the past, complicated rituals were developed for the harvesting of herbs: they were intended to ensure that the plants were gathered at a point when their therapeutic qualities were at their highest level. In essence, camomile should be harvested on a dry day in the summer when the flowers are fully open. Choose only clean and healthy flowers. Pick the stems with the flowers still attached.

Do not wash or rub the flowers, as you will damage them. Handle them as little as possible and dry or freeze them immediately in order to retain all their valuable properties. Remember that the more your hands or your equipment touch the flowers, the more volatile oil the flowers will leave behind and the less there will be left for your remedy, cosmetic or scented decoration.

To dry the flowers, remove their stems. The flowers are best dried spread out on clean, dry baking trays lined with kitchen paper and placed in a warm airing cupboard. Keep the door ajar to enhance air circulation and prevent rotting.

If you wish to retain the leaves and stems too, tie them in small bunches and hang them upside down in a warm, dry and well-ventilated room. Herbs are often seen drying this way in kitchens. This is not a good idea as kitchens are full of both moisture and fumes. When the flowers are dry, pack them into airtight plastic containers. Make sure you label them with both the type of herb and the date.

Flowers may also be frozen: simply place flower heads in plastic boxes or bags. Take care to label the container with both the plant and the date. Frozen flower heads retain their aroma longer than do dried ones.

BUYING CAMOMILE

If you do not choose to grow your own plants you may buy camomile flowers either fresh or dried from herbalists or mail-order companies (see page 78). Make sure that you are buying the variety you want; see pages 9–10 for the difference in character between Roman and German camomile species. Dried flowers keep for one year only. If you plan to make remedies, ensure that you are buying material suitable for internal consumption rather than inclusion in scented decorations.

USING HERBAL REMEDIES

Although camomile is one of the safest herbs used in medicinal remedies and cosmetic preparations, all herbs are toxic when they are taken in very large quantities. Taking large doses of camomile or making strong camomile tea has been known to cause nausea in some people, and some with sensitive skin may find that they have an allergic reaction if they touch fresh camomile.

In general terms, herbal remedies are safe and effective treatments for minor, everyday complaints. However, great care should be taken in pregnancy and with children and elderly people. Herbal remedies should not be used for ailments caused by your way of life, including exhaustion, stress or depression. In these cases, lifestyle changes are preferable to long-term use of herbal remedies. You should consult a medical, herbal or

Herbs and herbal preparations for sale at a market in the Cévennes, France.

homeopathic practitioner if you have a more serious complaint; in the case that a common ailment does not improve with herbal remedies; or to seek advice about your particular symptoms.

Take care not to mix camomile remedies with other herbal remedies or other medicine, as you may experience side-effects. If you are already taking some form of medication, seek professional advice before taking herbal remedies.

This book gives general information about how to use herbs to make herbal remedies and cosmetics. The author, editor and publisher cannot accept responsibility for side-effects caused by taking the herbal remedies discussed in this book.

CHAPTER 1

HISTORY AND MYTHOLOGY

Asclepius treating
a patient.

IT IS SO WELL
KNOWN EVERY
WHERE, THAT
IT IS BUT
LOST TIME
AND LABOUR
TO DESCRIBE IT.

CULPEPER,
HERBAL, 1653

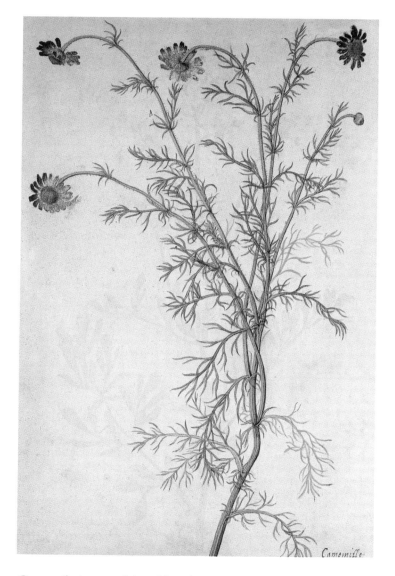

Illustration of
*Chamaemelum
nobile* by
J. le Moyne de
Morgues, c. 1568.

Camomile is one of the oldest documented herbs, and it is still
in widespread use today in much the same way as it was
thousands of years ago. Not only this, it has been one of the
most popular herbs used by people to heal a wide variety of
common ailments, to make cosmetics, and to enhance the living

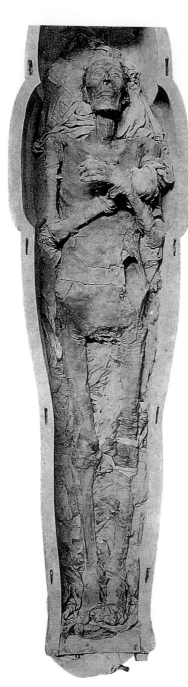

The mummified body of Ramses II.

environment – from the Egypt of the Pharaohs right up to the present day. Knowledge and practices of using camomile travelled from ancient Egypt to ancient Greece and Rome, and then to Europe where they were further spread as a result of the invention of the printing press, and the Renaissance.

ANCIENT EGYPT

The ancient Egyptians are the earliest documented culture to have understood and enjoyed the benefits of camomile. They revered it and dedicated it to their sun god. A papyrus document from approximately 2,800BC, during the reign of Khufu, records the use of camomile and other medicinal herbs for treatment of ailments such as skin disorders. *Matricaria recutita* (German camomile) was widely available in Egypt at that time, and is still cultivated on a large scale for both domestic use and export.

It is the Pharaoh Ramses II who has given us most information about the way camomile was used ritualistically. Ramses II, known as Ramses the Great, reigned for sixty-seven years, from 1279 to 1212BC, and died when he was probably in his nineties. At 183 cm (6 ft) tall, he towered over most of his contemporaries.

In 1881 his tomb was discovered near Deir el-Bahari. Archaeologists who analysed his mummified body found that his stomach contained camomile pollen and that his body was extremely rich in a coating of camomile oil. Egyptians believed that at death the soul left the body and that the practice of mummifying and burying the body would reunite the soul with

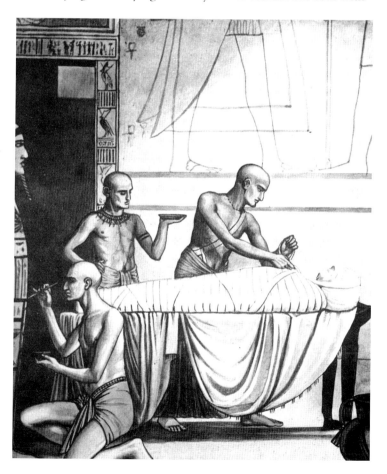

Egyptian embalmers bandaging a mummy while an assistant prepares the coffin.

the body and allow the mummy to live on in the afterlife. Ritual embalming was a complex procedure taking about seventy days, accompanied by much prayer and ceremony. The body of the dead person was removed to a 'place of purification', usually a tent set up as a workshop, where it was washed with Nile water. After this, an incision was made on the left side, and the liver, lungs, stomach and intestines were removed and placed in four Canopic jars (urns used specifically for this purpose). The organs' space in the body was filled with spices and resins. The brain was removed through the nostrils and the space filled with linen to preserve the features. The heart, considered the seat of intelligence, was left in the body. The nose of Ramses II had a distinctive profile and was packed with peppercorns and propped up with a small animal bone to preserve its shape. It was important to preserve the features, as it was believed that if they disintegrated, the *ka*, or personality, would disintegrate as well.

After this, Ramses II's body was preserved with precious oils and resins – including camomile oil. Amulets were placed on the body, which was then wrapped in linen bandages in a process taking about fifteen days. Each layer of bandages was brushed with liquid resin. The outer shrouds had the name of the Pharaoh written on them, and then the mummy was placed in a coffin. All the materials used were believed to have sprung from the tears shed by the gods at the death of Osiris, King of the Dead. Their use gave the dead man the power of these gods.

Once Ramses' body was in its coffin, a procession took it to its burial place. Servants followed bearing all the grave goods that the dead Pharaoh would need in the next life.

While the use of camomile was important in the ritualistic and religious treatment of the dead – and particularly the dead Pharaohs – it and other herbs also played a major role for the living in terms of remedies and cosmetics. The Egyptians were experts in the development and use of cosmetics, and were renowned, both in ancient times and now, for their use of herbal remedies. Indeed, many of their remedies have been passed on to other cultures, and have ultimately become major elements in modern medicines.

ANCIENT GREECE

Knowledge of and skills with herbs was passed from the ancient Egyptians to the ancient Greeks. The ancient Greeks used

Hippocrates, the father of modern medicine, recommended the use of herbal remedies.

camomile, which they called *kamai melon*, 'apple on the ground', because of the fresh apple scent it exudes when the plant is walked upon.

Hippocrates, the Greek physician dubbed the father of modern medicine, who lived from approximately 460 to 377BC, recommended herbal remedies, including camomile. As part of his medical treatments he prescribed perfumed fumigations. He is credited with the writing of fifty-three books about medicine, known as the *Corpus Hippocraticum*; these laid the foundations for the medical tradition of the West.

Breaking with the tradition of the ancient Egyptians, Hippocrates believed that the origins of disease were entirely natural – not caused by the gods. Hippocrates developed theories of how the human body was made up, what illness was and how it could be cured. He considered that everything in nature was made up of four elements: earth, air, fire and water. These elements were in harmony and influenced the seasons and the patterns of life.

Ancient civilizations found great use for herbal remedies.

They also governed the four humours that he considered made up the human body: phlegm, blood, black bile and yellow bile. Hippocrates believed that good health consisted of a proper balance in these four humours. If a person had one humour in excess, they became ill; the nature of their illness would determine which humour was the cause. The skill of the doctor lay in assisting nature to regain its balance by planning a programme for each individual, including herbal remedies, diet, lifestyle changes and exercise. The Hippocratic doctrine of humours offered a convincing explanation for the causes of disease, and was still in practice until the nineteenth century.

Knowledge and use of aromatic oils and herbal remedies also reached ancient Rome. Perfumes and aromatic oils were used

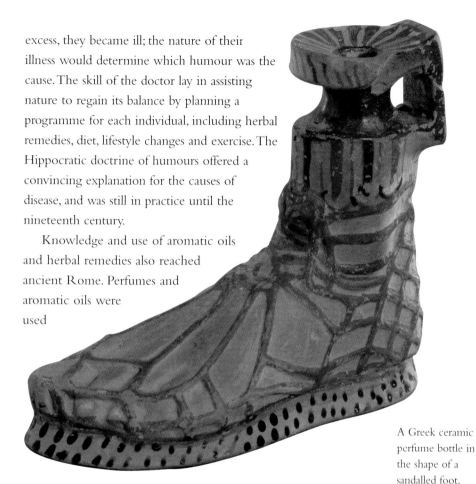

A Greek ceramic perfume bottle in the shape of a sandalled foot, dating from the seventh century BC.

extensively to anoint clothes and beds, and to perfume bodies and hair. The Roman Empire broadcast its ideals and practices across Europe.

INTO BRITAIN

Centuries later, in Anglo-Saxon Britain, herbal remedies were also used. The oldest surviving herbal, written in the vernacular – Old English – was published in the tenth century AD. Called *The Leech Book of Bald*, it consists of three collections of medical prescriptions. It lists eighty-eight remedies for particular ailments,

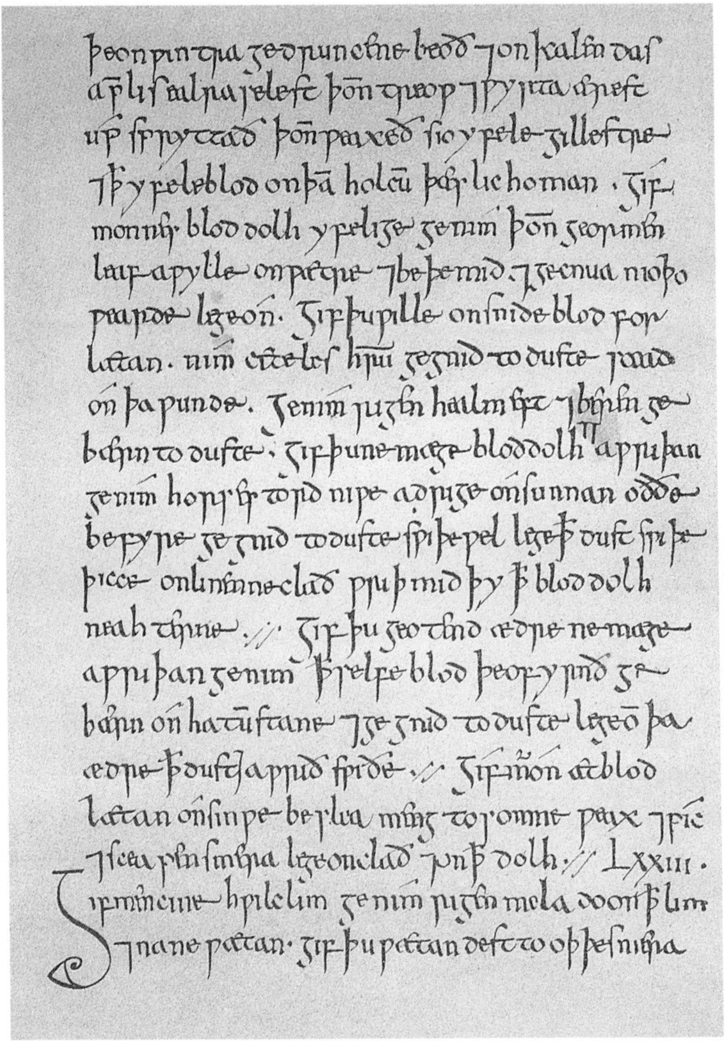

A page from
*The Leech
Book of Bald.*

describes the symptoms of various disorders and includes various herbal remedies sent to the English King Alfred from Elias, the Patriarch of Jerusalem. The book reveals a vast knowledge of native plants. As well as this, however, there is an element of magic, symptomatic of the understanding of illness at the time. According to this text:

'If any evil temptation come to a man, or elf or goblin, anoint his face with this salve, and put it on his eyes and where his body is sore, and cense him and frequently sign him with the cross; his condition will soon improve.'

It was centuries before medieval herbalists exploited the double-flowered cultivars they – and subsequent generations –

'Les Arts du Jardinage', a fifteenth-century miniature from *De Arte Memorie*.

A horticultural
scene from a
sixteenth-century
Calendar of
Months.

used to maximize the yield of flowers and oil for remedies. By
then, herbs were in widespread use, strewn over floors as
protection against plague and infectious diseases. Indeed,
camomile does have a valuable role to play in repelling insects.

THE GOLDEN AGE OF HERBALS

It was in the fifteenth century, with the invention of the printing
press, that the golden age of herbals dawned. Suddenly there was a

new and more efficient way to impart information to people. The sixteenth century, the age of Henry VIII and Elizabeth I of England, was marked by exploration and the discovery of new lands, plants and trade routes. This inspired many men of science, who were stimulated to know and understand the physical world around them. In this age many famous herbals were published, notably those by William Turner, John Parkinson and Nicholas Culpeper. They represented a huge advancement in the understanding of herbs and their therapeutic uses, made botanical illustrations into an art form as well as an invaluable tool in the identification and classification of herbs, and provided an invaluable legacy to scientists and herbalists in succeeding centuries.

A page dealing with camomile reproduced from William Turner's Herball.

WILLIAM TURNER

Dubbed the father of English botany, William Turner (c. 1508–68) was a Protestant, and therefore a believer in enquiry. He escaped religious persecution by fleeing to Italy where he studied medicine and botany. He was the author of the first books on botany to be published in English rather than the traditional language of scholarship: Latin. He published his famous *Herball* in three instalments between 1551 and 1568 with the full title of the first part as *A new Herball, wherin are conteyned the names of Herbes … with the properties degrees and naturall places of the same, gathered and made by Wylliam Turner, Physicion unto the Duke of Somersettes Grace*. It was dedicated to Elizabeth I, in whose favour he found himself. The book was illustrated and contained alphabetical listings of 238

plants native to Britain. In his work, Turner described camomile as 'very agreeing unto the nature of man, and ... good against weariness.'

JOHN PARKINSON

John Parkinson (1567–1650), Herbalist to Charles I, published two important works defining herbs and their medical uses. The first, *Paradisi in Sole Paradisus Terrestris. A Garden of all sorts of pleasant flowers which our English qyre will peermitt to be noursed up; ... together With the right orderinge planting and preserving of them and their uses and vertues,* appeared in 1629, in English. He went on to publish the largest herbal in the English language in 1640, *The Theatrum Botanicum: the theater of plants, or, an herball of a large extent.* In his work, Parkinson advocates the use of camomile baths to 'comfort and strengthen the soul and to ease pains in the diseased'.

NICHOLAS CULPEPER

Nicholas Culpeper (1616–54) has been immortalized for his

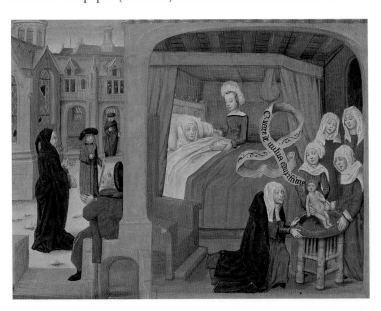

A mother and child being cared for in a fifteenth-century miniature *'Soins à la mère et à l'enfant'*.

work *The English Physitian: enlarged with 369 medicines made of English Herbs*, referred to as the *Herbal*, which was published in 1653 and which is still in print to this day. Culpeper got on the wrong side of the medical profession by translating the College of Physicians' Pharmacopoeia into English so that ordinary people would be able to read and understand it, and take responsibility for their own health and medical treatments by collecting herbs and making their own remedies free of charge.

An illustration of camomile from Culpeper's *Herbal*.

Culpeper refers to camomile as a gynaecological remedy to be taken against the after-effects of the 'careless midwife'. Indeed, the Latin name of German camomile, *Matricaria recutita*, refers to its role as a herb with gynaecological uses. His entry on camomile from his *Herbal* is represented here in its entirety. The contemporary understanding of the use of camomile is astonishing: it sums up so perfectly the benefits and widespread uses of the herb and echoes profoundly the information that is contained in modern books on herbs.

CAMOMILE

It is so well known every where, that it is but lost time and labour to describe it. The virtues thereof are as followeth:

29

Government and virtues. A decoction made of camomile, and drank, taketh away all pains and stitches in the side. The flowers of Camomile beaten, and made up into balls with oil, drive away all sorts of agues, if the part grieved be anointed with that oil, taken from the flowers, from the crown of the head to the sole of the foot, and afterwards laid to sweat in his bed, and that he sweats well. This is Nechessor, an Egyptian's medicine. It is profitable for all sorts of agues that come either from phlegm, or melancholy, or from an inflammation of the bowels, being applied when the humours causing them shall be concocted; and there is nothing more profitable to the sides and region of the liver and spleen than it. The bathing with a decoction of camomile taketh away weariness, easeth pains to what part of the body soever they be applied. It comforteth the sinews that are over-strained, mollifieth all swellings. It moderately comforteth all parts that have need of warmth, digesteth and dissolveth whatsoever hath need thereof, by a wonderful speedy property. It easeth all pains of the colic and stone, and all pains and torments of the belly, and gently provoketh urine. The flowers boiled in posset-drink provoke sweat, and help to expel all colds, aches and pains whatsoever, and is an excellent help to bring down women's courses. Syrup made of the juice of camomile, with the flowers in white wine, is a remedy against the jaundice and dropsy. The flowers boiled in lee, are good to wash the head, and comfort both it and the brain. The oil made of the flowers of camomile, is much used against all hard swellings, pains or aches, shrinking of the sinews, or cramps, or pains in the joints, or any other part of the body. Being used in clysters, it helps to dissolve the wind and pains in the belly; anointed also, it helpeth stitches and pains in the side.

Nechessor saith, the Egyptians dedicated it to the Sun, because it cured agues; and they were like enough to do it, for they were the arrantest apes in their religion I ever read of. Bachinus, Bena, and Lobel, commend the syrup made of the juice of it and sugar, taken inwardly, to be excellent for the spleen. Also this is certain, that it

Nicholas Culpeper occupied himself with the human body and astrology.

most wonderfully breaks the stone: some take it in syrup or decoction, others inject the juice of it into the bladder with a syringe. My opinion is, that the salt of it taken in half a drachm in the morning in a little white or rhenish wine, is better than either; that it is excellent for the stone, appears in this which I have seen tried,

31

viz. that a stone that hath been taken out of the body of a man being wrapped in camomile, will in time dissolve, and in a little time too.

Mathias de l'Obel (1538–1616), mentioned by Culpeper, was botanist and physician to Elizabeth I and James I. His magnum opus, *Stirpium adversaria nova*, published in 1570, was ground-breaking in its system of classifying plants. The genus *Lobelia* was named after him.

INTO THE TWENTY-FIRST CENTURY

The herbals of the Renaissance, including that of Culpeper, were in mainstream medical use up until the development of the modern drug industry in the nineteenth century. During the scientific revolution of the nineteenth century, chemists analysed and understood the constituents of herbs and their essential oils and were able to introduce new drugs based on extracted chemicals rather than the whole plant. While medical chemists confirmed the old wives' tales which passed down an understanding of the therapeutic uses of camomile, they also discovered more. Medicine passed into the hands of the professionals, and home-made herbal remedies, albeit based on thousands of years of practice, lost their credibility.

A book published in London in 1899, by Henry G. Greenish —*An Introduction to the Study of Materia Medica: Being a*

Abundant camomile flowers.

This traditional herbalist's shop in Amsterdam, Jacob Mooy & Co, was founded in 1743; with its old wooden drawers and shelves, it is the last of its kind in Holland.

Short Account of the More Important Crude Drugs of Vegetable and Animal Origin (Designed for Students of Pharmacy and Medicine) – credits the age-old traditional role of camomile by saying, 'The flowers have long enjoyed a wide reputation as a domestic

33

medicine.' However, he goes on to describe the plant with great botanical precision: 'The inflorescence of the wild plant is a capitulum [a rounded of flattened flower head] surrounded by two or three rows of overlapping bracts; the disc-florets are yellow, tubular, closely packed on an elongated conical receptacle, and surrounded by a single row of ray-florets with white ligulate corollas.' Information about the nature and use of camomile had passed into the hands of the scientist: 'When closely examined with a powerful lens, the lower part of the corolla may be seen to be sprinkled with minute, yellowish, shining oil-glands.'

At the cusp of the twenty-first century, however, there is a renewed desire to understand and harness the healing power of herbs in home-made recipes for remedies, cosmetics and scented decorations, and in making visits to traditional herbalists and homeopaths.

MYTHOLOGY

The power of herbs to bring relief to human suffering has made them a part of the mythology of different cultures and civilizations in various ages.

In Anglo-Saxon Britain, the Hippocratic theories of the composition of the human body and nature of disease were, for the moment, lost. To the people,

A barbed, flint arrow head was called elf shot by the Anglo Saxons.

34

illness was caused by such magical elements as 'elf shot' and 'flying venom'. (Elf shot was a name for prehistoric flint arrowheads found on the ground, which were believed to be used by elves to transmit illness.) To counter such ills, stronger magic was needed. They knew of camomile, which they called 'maythen', and accepted its healing powers although they didn't understand them. And they turned to Wodan, the god of the underworld, magic, inspiration, poetry and battle.

In the myths of the time, there was conflict between gods and monsters: the gods brought equilibrium to the human race and endowed it with art, wisdom and law, while the monsters sought to wreak havoc and bring chaos. Wodan was an important god in the mythology of northern Europe, a warrior god akin to the Roman god Mercury. He brought good luck in battle but finally condemned his followers to death and acted as a guide to the underworld. The other side to him stressed his ability to inspire poets and to divine the

An engraving of a wood elf, a magical bringer of disease.

future. In the mythology, camomile was one of nine sacred herbs given by Wodan to mortals to enhance their lives and bring them order out of chaos.

Opposite:
The mighty god Wodan seated on his throne.

CHAPTER 2
REMEDIES

THE BATHING WITH A DECOCTION

OF CAMOMILE TAKETH AWAY

WEARINESS, EASETH PAINS

TO WHAT PART OF THE BODY

SOEVER THEY BE APPLIED.

CULPEPER, *HERBAL*, 1653

A cup of
camomile tea
is pleasant
and soothing.

Peter Rabbit does not welcome his mother's ministrations!

'I am sorry to say that Peter was not very well during the evening.

His mother put him to bed, and made some camomile tea; and she gave a dose of it to Peter.

"One table-spoonful to be taken at bed-time." '

Beatrix Potter, *Peter Rabbit*

Camomile is among the most commonly used herbs in medicinal remedies. It has many traditional uses: for example, for the treatment of nervous tummy upsets, nausea, insomnia and skin irritation. And there are a variety of ways to exploit the

therapeutic qualities of the flowers and essential oil: from a relaxing tea to a soothing ointment or warming steam inhalation. This chapter gives many recipes for treating a wide range of common ailments.

PARTS USED
The flower; essential oil.

EFFECTS

• Anti-allergenic: relieves allergic symptoms caused by a sensitive response.

• Anti-emetic: helps prevent nausea and vomiting.

• Anti-histamine: helps avoid an allergic response by neutralizing histamine.

Camomile tea and a dish full of dried camomile flowers provide both a pleasant taste and ambience.

• Anti-inflammatory: reduces inflamed, swollen tissues, both internal and external.

• Anti-spasmodic: reduces involuntary spasms.

• Bitter: stimulates the production of digestive juices.

• Carminative: helps to relieve flatulence, indigestion and colic.

• Emmenagogue: induces or increases menstrual flow.

• Sedative: reduces tension, irritability, anxiety and headaches.

CONSTITUENTS
Coumarins, flavonoids (including rutin, which is believed to strengthen blood vessels); volatile oils (0.24–1.9 per cent) including chamazulene and other components that vary between German and Roman camomile; amino acids, anthemic acid,

40

choline, polysaccharide, plant and fatty acids, tannin and triterpene hydrocarbons.

Azulene, in both Roman and German camomile, contains most of the therapeutic qualities of the plant. It is the major constituent of the essential oil and gives the characteristic colour. It makes the oil soothing, pain relieving and an effective treatment for nervous conditions and skin irritation.

An illustration of camomile from the *Flore medicale*, published in Paris in 1821.

FORMS

Camomile may be taken in the form of preparations made from fresh flowers, dried flowers, essential oil, or the homeopathic preparation Chamomilla. Note that the essential oil is not to be taken internally.

Homeopathic Chamomilla is made from the whole plant when it is in flower. The plant is chopped and mashed to a pulp; the juice is extracted and is mixed with equal proportions of alcohol. It is then allowed to stand for eight days. After that, it will be diluted according to homeopathic formulae.

ESSENTIAL OIL

The essential oil of Roman camomile is pale blue, becoming pale yellow with age or exposure to sun. It has a sweet, fruity

perfume. The essential oil of German camomile is deep blue, turning green to yellow with age or sun exposure. It has a stronger scent that may be softened by combination with lavender or clary sage. The blueness comes from the chemical constituent azulene.

Volatile oil is found in plant tissues, in combination with other compounds. It is called essential oil when it has been extracted from the plant tissues and separated from the other constituents. Essential oils are a very potent and intense way of sampling the aromatic qualities of a plant, and, as they are toxic, should almost always be diluted in a 'carrier' oil such as almond oil. Essential oils can reach the human bloodstream in as little as 20 minutes after application to the skin. They are used in the treatment of muscle pains, digestive disorders and inflammations, and they are disinfectant and inhibit flies. They are not to be taken internally, and should not be used at all if you are pregnant (see page 14).

In 1930 a French chemist called René Maurice Gattefosse discovered that essential oils could be used for aromatherapy. The oils are rubbed into the skin, are readily absorbed and can treat physical disorders as well as emotional troubles. Camomile is used to treat patients experiencing anger.

BASIC RECIPES

Camomile is very versatile for the user, in as much as it may be taken both internally (for example, in the form of tisanes) or externally (as a massage oil or ointment). Here we give thirteen basic recipes which anyone can make. You will need a small amount of equipment:

- dark glass bottles
- glass jars with airtight lids
- small glass pots with airtight lids
- a fine-mesh sieve.

Opposite:
A few drops of camomile essential oil added to massage oil may be used to treat a range of complaints.

Always wash your hands thoroughly before starting work, and use clean and dry tools uncontaminated with foods or other herbs.

It is imperative to label and date any preparation that you make at home. Include the following information: the precise name of the herb (including the species), the part used, the type of preparation made and the date. For example: 'Roman camomile, fresh flowers, tincture, date'. Be sure to store all ingredients and preparations out of the reach of children.

INTERNAL APPLICATIONS

These three versatile treatments may be taken on their own or used as the base for another preparation. For example, a tisane may be drunk or poured into a bath. In general, when water is included in a recipe, use spring water. You may use dried camomile flowers or three times the amount of fresh flowers.

You may use dried camomile flowers (as here) or fresh in your remedies.

It is important not to make the recipes any stronger than is described here, not to take too much of the preparation, and not to mix preparations.

German camomile growing in a garden setting.

TISANE OR TEA

The word tisane is derived from the ancient Greek word for a medicinal brew. Also called an infusion, this is a simple and popular way of extracting the benefits from flowering and soft green plants. Camomile tea made from German camomile is the most popular herbal tea on the market and is widely drunk in Europe. Here is how to make your own.

Put 25 g (1 oz) of dried German camomile flowers (75 g/3 oz of fresh flowers) into a teapot. Pour on 500 ml (1 pint) of freshly boiled water and put the lid on the teapot to prevent the aromatic constituents from escaping. Leave to steep for 10 minutes. After this time, strain the liquid through a sieve into a jug, if you wish to drink it later, or straight into a cup for immediate consumption. You may sweeten the brew with a little honey if you wish. Keep the remainder in the fridge to be reheated or drunk cold as you choose. These quantities make three doses: a single day's supply.

For a single cup, use 1 heaped teaspoon of dried flowers to 1 cup of boiling water.

TINCTURE

A tincture uses alcohol to extract the medicinal properties of a herb. The alcohol acts as a preservative so that a tincture lasts for up to two years. In this way camomile can be used all year round, not just when the flowers are freshly harvested.

Put 900 g (2 lb) of fresh camomile flowers into a large jar with a tight-fitting lid. Add 1 litre (1¾ pints) of vodka and 500 ml (18 fl oz) of spring water. Put the lid on the jar and shake the jar. Store the jar in a cool, dark place for two weeks, giving it a shake occasionally. At the end of this time, strain the liquid through a sieve into a dark glass jar and store out of sunlight. Label and date the jar.

Uses for tincture are given in the following pages.

SYRUP

This is a useful preparation for treating colds and flu: see page 62.

Pour 500 ml (18 fl oz) of boiling water over 2 tablespoons of dried camomile flowers. Cool and strain the mixture. Then place it in a saucepan with 6 tablespoons of sugar. Heat the mixture gently until all the sugar has dissolved, then boil it until it thickens.

Inhaling the scent from a handful of dried flowers, including camomile, is in itself beneficial.

EXTERNAL APPLICATIONS

External applications of camomile tea or infused oil (including when added to bath water), or compresses and poultices using camomile flowers may cause allergic

reactions in sensitive individuals. This is most likely in people allergic to the *Asteraceae* (daisy) family.

Here are ten recipes for preparations that may be given externally. They include treatments that are rubbed in, inhaled, used for massage and placed directly on wounds.

INFUSED OIL

This recipe is used as a base for making ointments or massage oils. There are two methods of infusion: hot and cold. Both have the same effect and are used in the same way, but the hot-infused oil is ready sooner.

* For the cold-infused method, fill a large jar with camomile flowers, then cover it with an oil rich in essential fatty acids,

Liquid herbal preparations should be stored in dark glass bottles, such as these traditional blue ones, away from sunlight.

such as almond, olive or sunflower oil. Fit the lid, and leave the jar on a sunny windowsill for three weeks, shaking it daily. After this time, strain the oil, repeat the process with fresh camomile in the same infused oil, and leave the jar for a further two to three weeks. Strain the oil a second time and pour it into dark bottles with an airtight lid. Label and date the bottles.

- For the warm-infused method, place the camomile flowers in a bowl and cover them with oil. Place the bowl over a saucepan containing simmering water. Heat the mixture for one hour (taking care that the saucepan does not boil dry), then strain it and heat the infused oil again with fresh camomile. Strain it a second time and allow it to cool before bottling, labelling and dating it.

These methods have changed little since Culpeper's day:

The way of making them is this: having bruised the herbs or flowers you would make your oil of, put them into an earthen pot, and to two or three handfuls of them pour a pint of oil, cover the pot with a paper, set it in the sun about a fortnight or so, according as the sun is in hotness; then having armed it very well by the fire, press out the herb, &c. very hard in a press, and add as many more herbs to the same oil; bruise the herbs (I mean not the oil) in like manner, set them in the sun as before: the oftener you repeat this, the stronger the oil will be: at last, when you conceive it strong enough, boil both oil and herbs together, till the joice will be consumed, which you may know by its leaving its bubbling, and the herbs will be crisp; then strain it while it is hot, and keep it in a stone or glass vessel for your use.

Culpeper, *Herbal*, 1653

CAMOMILE STEAM INHALATION

This is a quick and easy way to treat sinusitis or hay fever.

Heat up 500 ml (18 fl oz) of camomile tisane (see page 45) until it is just boiling, and pour it into a bowl. Add 10 drops of camomile essential oil. Lean your face over the bowl, put a towel

A camomile steam inhalation soothes sinusitis and hay fever.

over your head and the bowl, close your eyes and inhale the steam deeply and with regular breaths for no more than 10 minutes. After this time, sit quietly in the room for about 15 minutes before moving away. This gives the mucous membranes time to adjust gradually to a lower temperature.

CAMOMILE CREAM

This recipe makes a gentle cream that is easily absorbed into the skin. It is used for dry and sore patches. Beeswax, anhydrous lanolin and glycerol may be obtained from pharmacists.

Combine 100 g (3½ oz) of almond oil or infused camomile oil, 25 g (1 oz) of white beeswax (grated) and 25 g (1 oz) of anhydrous lanolin in a bowl over a saucepan containing gently simmering water. Keep the water hot until the solids melt.

Add 25 ml (1 oz) of glycerol and 60 g (2¼ oz) of dried camomile flowers, and heat the mixture for three hours (taking care that the saucepan does not boil dry). Sieve the mixture into a bowl and stir it continuously until it has cooled. After this, store the cream in airtight, dark jars in the fridge. It keeps for about two months.

Camomile flowers can be steeped in mixtures containing beeswax, almond oil and lanolin to make a gentle and effective cream.

CAMOMILE OINTMENT

The following recipe makes an ointment or a base for massage oils.

Melt 25 g (1 oz) of beeswax and 25 g (1 oz) of anhydrous lanolin in a bowl over a saucepan containing gently simmering water. Add 100 ml (3½ fl oz) of infused oil (see page 48). Pour the ointment into dark glass jars and allow it to cool. Fit airtight lids on to the jars. Label and date the jars.

MASSAGE OIL

Place 20 drops of Roman camomile essential oil (or a combination of lavender, clary sage and German camomile essential oils) in a dark bottle. Add 50 ml (2 fl oz) of jojoba oil. Put the lid on and then roll the bottle gently between the hands to mix the ingredients. Never shake the bottle. Label and date the bottle and store it in a cool, dark place.

Jojoba oil is almost odourless and makes the skin feel as smooth as satin. Alternatively, you could use:

- almond oil: very popular in massage oils and almost odourless
- olive oil: calming and pungent
- hazelnut oil: nourishing, stimulating and pungent.

RELAXING CAMOMILE RUBS

For stiffness in joints or muscles dilute a few drops of camomile essential oil in 20 ml (³/₄ fl oz) of hazelnut or almond oil.

To treat stomach cramp, perhaps associated with menstruation, mix 20 ml (³/₄ fl oz) of hazelnut oil with 25 ml (1 fl oz) of cramp bark tincture (made from *Viburnum opulus*, the bark of the guelder rose, a sedative herb that soothes the uterus) and 40 drops of camomile essential oil.

CAMOMILE BATH

Add up to 5 drops of Roman camomile essential oil to your bath water for a soothing bath: the oil of this species is non-irritating and non-sensitizing. Instead of essential oil, you may add 500 ml (18 fl oz) of camomile-infused oil for the same effect (but see the warning on page 47).

Alternatively, gather a handful of camomile leaves and flowers, pop them into a square of muslin, tie the corners of the muslin, and swish it through the bath.

CAMOMILE POULTICE

This is a pulp made from camomile flowers used for relieving minor wounds and burns.

To make a poultice place a handful of camomile flowers in a bowl of boiling water until they have softened. Allow them to cool slightly then put them between two layers of gauze and apply to the affected part.

COOL CAMOMILE COMPRESS

For relief from inflammation, make some camomile tea at normal strength or mix one part camomile tincture with four parts water.

Allow the liquid to cool then make a cool compress by soaking a clean cloth in the mixture – a swab of cotton wool for a small area or a clean flannel for a larger area – and wrap it around the part that is painful.

HOT CAMOMILE COMPRESS

This is a warming and comforting treatment for boils, rheumatism, toothache and backache.

Fill a bowl with just-boiled water and add 4 drops of camomile essential oil. Stir to mix. Soak a clean flannel in the mixture, wring it out and place it on the affected part until it has cooled to blood heat.

Repeat as many times as you feel necessary or until the water in the bowl has cooled.

Camomile is one of the safest herbs for breastfeeding mothers and their babies. A mother and baby enjoy each other's company in this painting of *Motherly Love* by Josef Danhauser, 1839.

PREGNANCY, BREASTFEEDING AND BABIES

Particular care needs to be taken with herbal remedies if you are pregnant, breastfeeding or responsible for a poorly baby or young child. Here we describe in depth how to use camomile safely in these conditions.

PREGNANCY

Avoid camomile essential oil in pregnancy. It acts as an emmenagogue: that is, it brings on bleeding. If you are suffering from morning sickness, take a cup of Roman camomile tea or a few drops of tincture in bed before rising. Be aware, however, that excessive use of either species of camomile is not recommended during pregnancy.

Traditional recipes exist to relax a woman in labour. A relaxing tea may be made from camomile, linden flowers and lemon balm. Massage has an invaluable role to play in labour: in advance, make up a mixture of camomile, lavender, clary sage and clove oils in almond oil.

Raspberry leaf is well known among midwives for its ability to tone the womb in the last few weeks of pregnancy. A good postnatal tea mixes camomile and raspberry leaf to relax the new mother and help her womb contract to its normal size.

BREASTFEEDING

Camomile tea is considered safe for breastfeeding mothers when used in moderation; however, drinking it in excess is not recommended. Taking camomile tea up to three times a day is fine, but taking camomile internally to a greater extent than that may interfere with the production of breast milk and cause side-effects in both mother and child. It may also bring back menstruation in a nursing mother.

If you are experiencing sore nipples as a result of breastfeeding, apply camomile cream to them after feeding.

Many generations of mothers have believed that a recipe for a tisane called 'mothers' milk tea' may increase milk supply. The ingredients are: camomile flowers, fennel seeds, coriander, lemongrass, borage leaves, blessed thistle leaves, star anise, comfrey leaves and fenugreek seeds. However, it should be emphasized that if milk supply is low, the best ways to increase it

are to eat nourishing food, take plenty of rest and feed your baby often, including during the night.

BABIES AND CHILDREN

Camomile is one of the safest herbs for babies and children; in Germany camomile and other herbal teas are widely given even to infants. However, the La Leche League, the

Camomile tea may be drunk in moderation by breastfeeding mothers and be given to babies over six months old.

international organization that offers support and information to breastfeeding mothers, stresses that camomile tea should be considered a drug and should not be given to babies of less than six months. If symptoms persist or worsen, consult your doctor immediately. Take particular care not to overdose your child: see the recommended doses for children below. Also take care not to give your child several different treatments at the same time.

Colic and teething

An older baby or child may be given a weak camomile tea to drink (see children's doses, page 60) if they have colic or a sore tummy, or are teething. Alternatively, pour camomile tea into the bath water or apply a cold compress of camomile to the child's tummy (see page 54). Be sure to check in advance whether your baby or child is allergic to the plant.

Homeopathic Chamomilla 3x may also be given to babies for colic and teething, but you should seek advice from a qualified homeopath first.

Conjunctivitis

Bathe the child's eye in a very weak camomile tea mixture. Be sure to assess first whether your child has an allergic response to the plant. If symptoms continue be sure to seek advice from your health professional.

High temperature

A young child with a high temperature deriving from a common self-limiting infection such as an ear infection may be treated at home to bring down their temperature. However, if in doubt, or if symptoms persist, be sure to seek advice from your health professional. In general, a child should be allowed to sweat out the fever; bringing down the temperature with

A child with a fever may be soothed with a little camomile tea and refreshed with a weak solution of camomile.

commercial preparations is not the best policy. Camomile can help children feel more comfortable while they are feverish: camomile tea sweetened with a little honey calms them, while sponging with a flannel soaked in a bowl of warm water to which a drop of camomile essential oil has been added makes them feel a little fresher.

Nappy rash

Camomile cream applied to babies' bottoms is a gentle treatment for nappy rash. Giving a camomile bath with either essential oil or camomile tea also helps soothe sore skin, but again check for sensitivity to external application of the tea.

Sleeplessness

A breastfeeding mother may ease tension and sleeplessness in her infant by taking a cup of camomile tea herself; many of the benefits of the tea are passed to her child through her breast milk. Again, this should be undertaken with caution. A way to give relaxation and pleasure to both mother and child is to massage the baby with a mixture of 20 ml ($^3/_4$ fl oz) of sweet almond oil and 1 drop of camomile essential oil.

RECOMMENDED DOSES FOR CHILDREN

6 months–1 year	5% of adult dose
1–3	10%
3–5	20%
5–7	30%
7–9	40%
9–11	50%
11–13	60%
13–15	80%
15 upwards	100%

GENERAL AILMENTS

Camomile treats with success a wide range of disorders including nervous digestion, insomnia, wounds, sunburn and painful menstruation. Here is a list of common ailments where camomile can help, together with an explanation of how to treat them using the recipes earlier in this chapter.

It should be stressed that for major or persistent problems you

should consult a qualified
herbalist, homeopath, or your
qualified medical practitioner.

Anxiety

Camomile is very effective for
calming general anxieties. There
are various treatments possible:
taking a camomile bath, drinking
camomile tea, or having a
relaxing massage with camomile
oil (see page 52). You may also
choose to perfume a room in
your home or your office with
fragrant water or pot-pourri
(see page 75). If your anxiety
becomes debilitating or long
term, seek professional help.

Asthma

In the case of mild asthma
attacks take a camomile steam
inhalation (see page 50). If you
are suffering a more major attack, consult your qualified medical
practitioner at once.

Camomile
essential oil
is poured into
a vaporizer,
and may give
relief from
colds and sinusitis.

Boils

To relieve the pressure and pain and reduce the infection of
boils, apply a hot compress (see page 54) of diluted camomile
essential oil.

Burns

To soothe burns or sunburn soak a pad of cotton wool in cool camomile tea and place it over the affected part. See also page 67 for a recipe for after-sun lotion. For more severe burns consult your medical practitioner.

Colds
Take 2 teaspoons of camomile syrup (see page 46) to clear the nose and aid restful sleep. Alternatively, you may add 3 drops of camomile essential oil to a bowl of warm water and leave it in the bedroom overnight.

Cramp
To ease the pain of cramp use the special relaxing rub recipe: see page 52.

Diarrhoea caused by stress
To calm the stomach and ease the stress that can cause diarrhoea take three cups of camomile tea daily. If symptoms persist, consult a professional.

Eczema
Camomile cream, gently applied to patches of eczema, helps alleviate the irritating sensitivity. A bath with camomile essential oil also helps.

Hay fever
For effective relief from hay fever, place a drop each of camomile essential oil and lemon essential oil on a tissue and inhale. Alternatively, take a camomile steam inhalation (see page 50).

Indigestion
Take a cup of camomile tea after meals to aid the digestion and calm the stomach.

Insect bites

To relieve the irritation caused by insect bites, rub a drop of camomile essential oil into the skin surrounding the bite.

Insomnia and nightmares

Take a cup of camomile tea at bedtime, sweetened with a little honey; try adding a small amount of sage to the mixture for a relaxing bedtime drink. A warm camomile bath before bedtime should also help you to relax.

Migraine

In the case of a severe headache or migraine apply the camomile relaxing rub to the temples.

Mouth sores

Gargle with camomile tea to reduce inflammations of the mouth.

Nausea

Take camomile tea three times a day to combat all forms of nausea, including travel sickness and morning sickness. If you are pregnant, see 'Pregnancy, breastfeeding and babies', page 55.

Nervous tummy upset

Camomile is particularly effective at calming nervous stomach upsets: camomile tea, tincture, or compress (see page 54) are all gentle ways to alleviate the problem.

Period pain

Take a camomile bath to ease the cramps associated with menstruation. You may also find it helpful to squat over a camomile inhalation.

A massage with camomile essential oil brings relief and relaxation to aching joints.

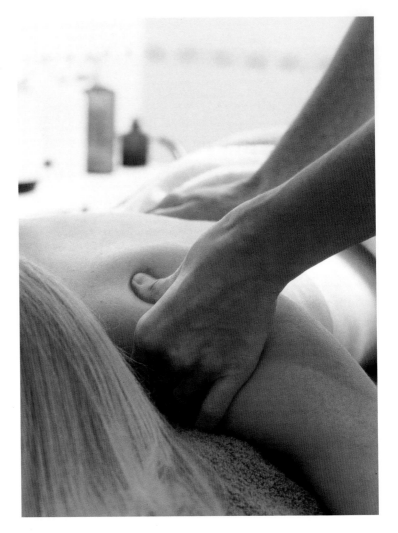

Sinusitis

A camomile steam inhalation with added essential oil of eucalyptus helps to ease the pressure and pain of sinusitis.

Stiffness

If you have a stiff shoulder or knee, apply a relaxing camomile rub (see page 52) to the affected part.

Stye

Place used camomile teabags (cooled but still warm) over the affected eye.

Travel sickness

Take a cup of camomile tea before travelling. In the case of children, see children's dosages, page 60.

Varicose veins

For effective relief from varicose veins, soak a clean flannel in cooled camomile tea and wrap it around the affected part.

Wounds

Camomile cream applied to minor wounds helps soothe and mend the injury. In cases of more major wounds or continued pain, however, you should seek medical advice.

CHAPTER 3
COSMETICS

Lady Hera went in and closed the polished doors behind her. She began by removing every stain from her comely body with ambrosia, and anointing herself with the delicious oil she uses. It was perfumed and had only to be stirred in the Palace of the Bronze Floor for its scent to spread through heaven and earth. With this she rubbed her lovely skin; then she combed her hair, and with her own hands plaited her shining locks and let them fall in their divine beauty from her immortal head. Next she put on a fragrant robe of delicate material that Athene with her skilful hands had made for her and lavishly embroidered.

Homer, Iliad, Book XIV

Camomile has been a valued ingredient in cosmetics since the days of the Pharaohs. Anointing the body, hair and clothes with aromatic oils – particularly those with mystical or religious significance – was an important ritual in ancient cultures, and people became skilled at producing luxurious cosmetics for use by the elevated classes.

In our day, the flowers and essential oil of camomile are still widely used in the commercial production of cosmetics, soap, hair products, bath products and sophisticated perfumes. This chapter gives details of cosmetics that you can make at home, quite simply, so that you can enjoy both making and then using the preparation.

MAKING YOUR OWN COSMETICS

Camomile has traditionally been used to lighten and condition hair and to make an anti-allergenic skin cream. Be aware that sensitive skins or susceptible individuals may experience an adverse allergic reaction, so always do a patch test before using your home-made cosmetic.

FACE

Camomile is extremely versatile and can be used to make gentle cleansers, soothing creams, or a refreshing spritzer.

AFTER-SUN LOTION

For a gentle, cooling and restorative after-sun lotion blend 1 tablespoon of jojoba oil with the following essential oils: 3 drops of camomile, 2 drops of lavender and 1 drop of peppermint. If your skin feels raw after exposure to the sun, put 4 drops of camomile essential oil and 3 drops of lavender essential oil into a bath.

Opposite:
This relief from a Greek vase shows a servant perfuming her mistress's hair in the presence of the winged god of love, Eros.

CLEANSER

For a deep, gentle cleanser, take a steam inhalation with a few drops of camomile essential oil.

COOLING WASH

For a cooling wash, add a few drops of camomile essential oil to a basinful of water and wash face and hands as normal.

CREAM FOR DRY SKIN

A cooling face wash refreshes the skin.

Opposite: Camomile may be used in the preparation of a range of gentle cosmetics and perfumes.

For dry or sensitive skin, rub camomile cream into the affected part. In the case of badly affected areas, you may dab a drop of camomile essential oil only on to the skin and rub it in. As it is very rapidly absorbed, you may need to apply a second drop. Please note that essential oil should not be applied if you are pregnant.

EYE SOOTHER

For tired and swollen eyes make a cup of camomile tea from tea bags; then, when cool, place the used tea bags over your eyes. Alternatively, bathe the eye in an eyebath containing carefully strained camomile tea.

FACE SPRITZER

For a cooling spritzer for the face, spray refrigerated camomile tea on to the skin. This reduces inflammation and refreshes the complexion.

HAIR

One of the traditional uses of camomile is to lighten blonde hair. Its gentle action and fresh perfume make it suitable for all hair-care preparations.

Opposite:
Camomile has traditionally been used to lighten blonde hair.

HIGHLIGHTER

Make up an infusion with 7 g (¼ oz) of dried camomile flowers in 100 ml (3 ½ fl oz) of boiling water. Leave it to stand for 20 minutes before straining it. Pour it over the hair and rub it in. Leave it for 20 minutes before rinsing it out.

CONDITIONER

For a gentle conditioner use 100 g (3 ½ oz) of coconut oil. It is either liquid or solid in form, depending on the ambient temperature. If solid, you may need to melt it in a bowl over a saucepan of gently simmering water. Stir in 20 drops of camomile essential oil. Apply it to wet hair before shampooing, and massage it in. Leave it for up to 2 hours before shampooing it out.

SHAMPOO

To make camomile shampoo mix 4 drops of camomile essential oil with 1 tablespoon of unscented baby shampoo. Wash your hair as normal.

CHAPTER 4

SCENTED DECORATIONS

CHAMOMYLLE ... IS

VERY AGREEING UNTO

THE NATURE OF MAN, AND ...

IS GOOD AGAINST WEARINESS ...

WILLIAM TURNER

Camomile has, for hundreds of years, been brought into the home so that its gentle aroma may fill the rooms in order to bring a calm atmosphere and repel insects and infectious diseases. In medieval times the floors of rooms would be strewn with herbs to foster a healthy ambience. The art of making scented decorations reached its zenith in Elizabethan days when the mistress of the house would gather aromatic herbs and flowers and dry them in the still room before making pot-pourri, scented cushions, medicines and cosmetics. Making these kinds of scented decoration was considered just as valuable a way of harnessing the herbs' therapeutic and aromatic qualities as was making medicines, and was just as much a responsibility. Receipt books – containing recipes for remedies, cosmetics and scented decorations – were passed down from mother to daughter over generations.

MAKING YOUR OWN SCENTED GIFTS

Scented decorations are simple to make, and those that are wrapped in fabric require just as much, or as little, sewing as you are prepared to do. If you do get out your sewing machine, ensure that you leave one edge of the fabric easy to undo so that you can open it up to revitalize the camomile flowers with a drop of essential oil after a couple of months.

SCENTED SACHETS

To make scented clothes sachets for hanging in a wardrobe or putting in a drawer, place dried camomile flowers in a square of fabric of your choice. Gather the corners together and tie them with attractive ribbon or lace. Leave a loop by which to hang it if you choose to place it over a clothes hanger in the wardrobe.

Opposite:
An illustration of camomile from the medieval text *Hours of Ann of Burgundy.*

FRAGRANT CAMOMILE SPRAY

For a fragrant room spray pour 500 ml (18 fl oz) of boiling water onto 12 tablespoons of fresh camomile flowers. Cool and strain the mixture. Then mix it with 150 ml (5 fl oz) of vodka. Add 10 drops of camomile or rose essential oil. Store it in a dark glass bottle, clearly labelled and dated.

Opposite: Candles can be scented with the fragrance of camomile.

POT-POURRI

Camomile has a valued place in any pot-pourri made from scented flowers and herbs, including lavender, rose petals, orange blossom, peony petals, marjoram and oregano.

Choose a variety of dried ingredients and, for every 4 cups (900 ml/32 fl oz) volume of mixture, add 1 tablespoon of orris root which will fix the perfume, as well as 5 drops of essential oil of your choice. Place the pot-pourri in an airtight container and store it in a warm, dark place such as an airing cupboard for six weeks, shaking it occasionally.

A pot-pourri stall in a market in St Tropez, France.

SLEEP PILLOW

For a simple sleep pillow to aid restful sleep and lull you back to sleep should you awaken, tie camomile flowers and rose petals in a large handkerchief and tuck it under your pillow. For a more ambitious project, sew a pretty case for your herbs and tuck that under your pillow.

SCENTED CUSHION

Make a cushion cover out of the fabric of your choice, leaving one side unstitched. Half fill the cushion with kapok filling. Then add a bag of camomile pot-pourri (see page 77), add more kapok filling, and sew shut the last side.

BATH SCENTER

Gather fresh or dried camomile and lavender flowers into a square of muslin. Tie the muslin with a ribbon and loop it over the hot tap so that the warm water runs through the bag.

SCENTED CANDLE

A scented candle may be made in several ways. Most simply, light a wide candle and let the top of the wax melt. Put out the flame and mix a few drops of camomile essential oil into the melted wax. The fragrance should last for an hour.

Alternatively, you may buy a candle-making kit and add a few drops of camomile essential oil to the mixture just before pouring it into the candle mould.

Opposite:
Floating candles may be perfumed with camomile essential oil to exude a relaxing aroma.

FURTHER READING

Arber, A., *Herbals: Their Origin and Evolution; A Chapter in the History of Botany, 1470–1670*, Cambridge University Press, Cambridge, 1986

Bown, D., *The Royal Horticultural Society Encyclopedia of Herbs and Their Uses*, Dorling Kindersley, London, 1995

Bremness, L., *Crabtree & Evelyn Fragrant Herbal: Enhancing Your Life with Aromatic Herbs and Essential Oils*, Quadrille Publishing, London, 1998

Culpeper, N., *Culpeper's Complete Herbal & English Physician*, Parkgate Books, London, 1987

Hillier, M., *The Little Scented Library: Sachets and Cushions*, Dorling Kindersley, London, 1992

Lawless, J., *The Illustrated Encyclopedia of Essential Oils: The Complete Guide to The Use of Oils in Aromatherapy and Herbalism*, Element, Shaftesbury, 1995

Nice, J., *Herbal Remedies for Healing: A Complete A–Z of Ailments and Treatments*, Piatkus, London, 1990

Ody, P., *Simple Healing with Herbs: Herbal Treatments for More Than 100 Common Ailments*, Hamlyn, London, 1999

Shaw, N., *Herbalism: An Illustrated Guide*, Element, Shaftesbury, 1998

Chambers's Encyclopaedia, Chambers, London, 1908

USEFUL ADDRESSES

G Baldwin and Co. – supplier of herbs and essential oils by post
173 Walworth Road
London SE17 1RW
Tel: 0171 703 5550

Neal's Yard has a mail-order service:
Tel: 0171 627 1949

National Institute of Medical Herbalists
56 Longbrook Street
Exeter EX4 6AH
Tel: 01392 42602230

La Leche League
PO Box 29
West Bridgeford, Notts
NG2 7NP

ACKNOWLEDGEMENTS

Thanks to Collins & Brown for permission to reproduce text from Culpeper's *Herbal*.

Thanks go to Deni Bown, for her comments and expertise, to Frances Vargo and Margot Richardson for their invaluable contributions to the book, and to my father, Brian Ferry, as well as Rachel al-Azzawi and Nina Payne for their help and encouragement. Particular thanks are due to Richard, who provided me with great support and opportunities to write in such as way that Matilda didn't realize Mummy was busy.

PICTURE ACKNOWLEDGEMENTS

The publishers would like to thank the following sources for their kind permission to reproduce the pictures in this book:

A-Z Botanical/Bob Gibbons 6, Lino Pastorelli 10t; **Ancient Art & Architecture** /Ronald Sheridan 21, 26; **Gillian Beckett** 9, 10b; **The Bridgeman Art Library, London/**, Ashmolean Museum, Oxford UK, Perfume bottle in the shape of a sandled foot, Greek, from Taranto 7th Century BC (ceramic) 23 , Musée National du Moyen Age et des Thermes de Cluny, Paris/Peter Willi, Woman Bathing surrounded by Attendants, from a series of tapestries depicting the Noble or Courtly Life, 16th century(tapestry) 53, Victoria & Albert Museum, London UK, Alhoni Camomile: (Chamaemelon nobile),c. 1568, by J le Moyne de Murgues (d.1588) 17; ©**Carlton Books Ltd** 42, 51, 69; **Cephas** 38, Michael Barberousse 40, 57; **Jean Loup Charmet** 25, 28, 35, 41, 66; **Corbis** /Bettmann 19, Gianni Dagli Orti 16, 22, Christel Gersetnberg 36, Ali Meyer 55, Adam Woolfitt 33; **Courtesy Culpeper Limited** 29, 31; **ET Archive** 34, 72; **Garden Matters** 11, 15; **Garden Picture Library**/Linda Burgess 12, Michael Howes 13, Joanna Pavia 45, Laslo Puskas 32; **John Glover** 7, 8; **Robert Harding Picture Library**/Michael Leis/Freudin ©Burda 50 **The Image Bank**/Laurie Rubin 76; **Retna Pictures Ltd.**/P. Reeson 6; **Tony Stone Images**/Christopher Bissell 64, James Darrell 70, Carol Ford 47, Luc Hautecoeur 74, Tony Hutchings 68, David Jacobs 75, Vincent Oliver 59 Christel Rosenfeld 44; **Illustration from THE TALE OF PETER RABBIT by Beatrix Potter. Copyright© Frederick Warne & Co.1902,1987. Reproduced by kind permission of Frederick Warne & Co. 39**

Every effort has been made to acknowledge correctly and contact the source and/copyright holder of each picture, and Carlton Books Limited apologises for any unintentional errors or omissions which will be corrected in future editions of this book.